Introduction

On the eve of her first double digit birthday, Ella went to bed thinking about her mother's last words, "Oh, Ella, sometimes I wish I still had the little girl who enjoyed fairy tales" and dreaming about her favorite thing: Math! Math! and more Math! This time, however, her math symbols, formulas, vocabulary, and all things math were mixing with all things fairy tale!! Could she still make sense of her world in a make-believe world? These are her adventures as she moves from story to story.

Once upon a time in a land not so far away lived a girl name Ella, but everyone called her Mathella because she loved everything and anything that has to do with math. This morning, there was a cool, crisp feeling in the air. Ella just loved the fall. Today would be the perfect opportunity to wear her new, red cape.

Before she could get out of the bed, her dog, **Integer**, came bouncing **up and down** on her bed. She would have to take him on a walk. Oh well, she might as well walk him to Grandma's house.

As Ella got dressed, she realized that **Integer** always made her smile. She was **positively** not going to allow any **negativity** into her life today. Plus, her cape matched **Integer's** leash perfectly.

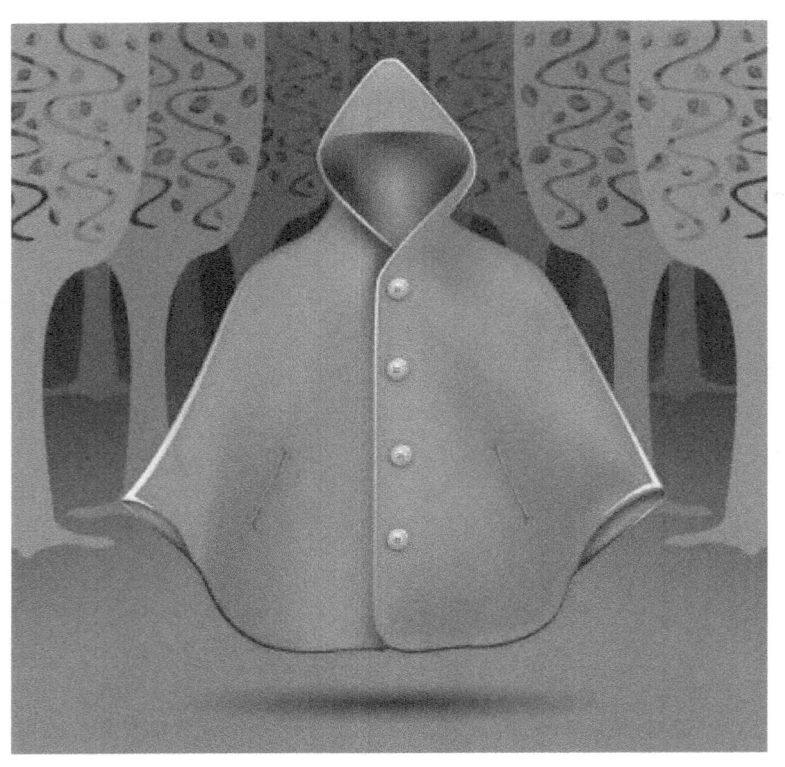

Ella **absolute**ly knew that the **distance** from her house to Grandma's house was 120 yards, about the length of a football field.

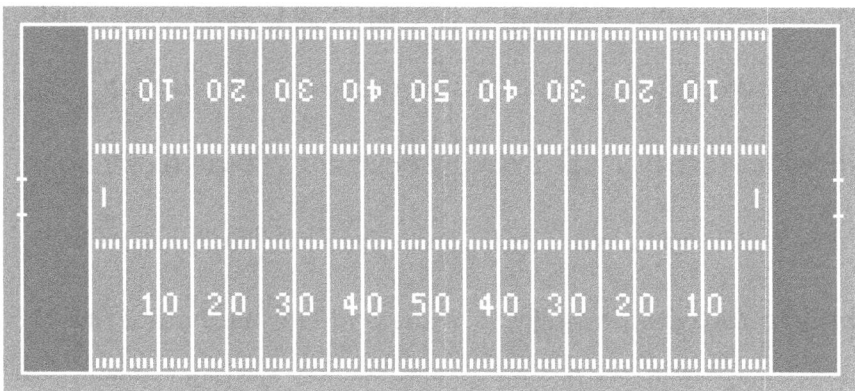

If she went right, she would **gain** more time. If she went left, she would **lose** more time. As she headed out the door, Mother handed her a basket full of goodies for Grandma, who hadn't been feeling well for the past week. Ella couldn't wait to get to Grandma's house. Grandma, after all, was one of her best friends.

She had only gone 4.75 yards when Integer ran **back** 7.89 yards to fetch his ball. Of course, this went past her house and onto the neighbor's front yard. Good thing she and the neighbors were friends. She would now need to recalculate the distance to Grandma's house. The first thing she did was **add the inverse** of the distance from the neighbor's house to her house (which is, incidentally, her favorite dessert). This brought her back to **zero (home again).** She then added that number to 120 yards. Ella could tell that walking Integer would not be as easy as she thought.

With that ordeal behind her, Ella headed towards Grandma's house with only **positive** thinking. This time, she heard strange noises. Must be my imagination, running away with me, Ella thought. But when Integer started barking, she knew he had a **rational** reason. He would either **repeat** or **terminate** his barking, but he always did a **fraction** of these. It was very unlike him to act irrationally. Suddenly, a strange wolf-like creature bounded out of the woods. Ella would have screamed, but her cape (or math) made her feel like a superwoman!

The wolf-like creature engaged in a math battle with Ella (who would have thought that wolves liked math). "If I **subtract a negative** from your life, could you **positively answer** the question?" Without missing a beat, Ella sprang into a song and dance routine (dancing really was her favorite thing to do).

"When you **add** the same steps forward and the same steps back, they form a zero cuz opposites attract

And you know, when one sign is higher and the other low

Don't add, subtract to find which way to go

Take the bigger and choose that sign

You'll get it right if you do it every time

If the signs the same, then just use that sign

Add them together, the sum will be just fine

If you can't do that in record time

Please feel free to use a number line."

The wolf should have been impressed, but alas he wasn't. Ella had not answered his question. She had been somewhat distracted by Integer running up and down as she sang. He was doing a great job of following the directions in the song (the money and time she had spent training him had not been wasted).

As the creature crept toward Ella, the only thing she could think of was math, math, and more math! Integer, on the other hand, knew a little something about **removing** things. When he would steal the food (which was **positive**ly delicious) in the kitchen, there was **less** of it (this action would cause him to get in trouble). But when he would remove the trash, he got rewarded for his behavior with a nice, big bone. He learned that **subtracting a negative** caused a **positive** reaction.

This walk was not turning out as planned. With Integer's running and the wolf's questioning, Ella still had 62.6 yards to go. At least the wolf was nowhere in sight! She finally arrived at Grandma's. Before she could ring the doorbell, the door opened, and a hand stuck out. Grandma had been sick, but her hands had never been skinny or hairy. The cunning wolf had gone to Grandma's house. However, he was no match for Ella. She was not nicknamed Mathella for nothing. Again, her math skills would save her for the day.

She approached the wolf with poise and confidence (one of the many benefits of learning math). Her **whole** day had been filled with **positive** and **negative** things. A **fraction** of it was spent calculating yards using **decimals.** It was only **rational** that she was not in the mood to reason with a wolf. She started attacking him with math problems (the only way to effectively attack a wolf). He got so frustrated that he ran out of the house yelling something about **integers** and **rules**. She even heard him mention multiplication and division, but that was for another day. She still hadn't seen Grandma. Just then, Integer walked in with Grandma. After a long hug, they sat at the kitchen table, opened the basket, and ate the goodies. It was another productive day in Ella's world.

Math Worksheets

The story used context clues to give the definition of math vocabulary. First, use the context clues to match the words. Then use your math textbook or other resources to write the definitions.

Matching: match the words with the correct definition

1. Integers   A. distance
2. Absolute value B. sum of opposites number
3. Additive inverse C. positive and negative numbers

Definitions: Use textbook/resources to give the mathematical definition

4. Integers_____

5. Absolute Value_____

6. Additive Inverse_____

In the story, Ella had only gone 4.75 yards when Integer ran back 7.89 yards to fetch his ball.

    7. How many yards did Ella lose?
    8. How many yards does she need to walk to get back home?
    9. What is the new total yards to Grandma's house?

In the story, the wolf asked Ella about subtracting integers. However, her song was about **adding integers**. Use line(s) from the song to answer the following questions:

    10.   When does the sum of two numbers equal zero?

    11.   What math operation is used when adding different signs?

    12.   How do you add the sum when the signs are the same?

Ella suggested that a number line could be used. Draw a number line to model the sums:

13. 6 + (- 6)

14. -10 + 7

15. 7 + 7

16. -2 + (-8)

17. In the story, Integer had to help Ella with the wolf's question about subtracting. What did he notice about **subtracting a negative**?

18. Rewrite the following numbers using Integer's observation: - - (-6)= _____ -(-4.5)=_____ -(-1/5)= _____

Use the following paragraph to show examples of rational numbers

Her **whole** day had been filled with **positive** and **negative** things. A **fraction** of it was spent calculating yards using **decimals.** It was only **rational** that she was not in the mood to reason with a wolf.

The last thing Ella did was attack the wolf with math problems. Solve the math problems that she used (http://www.mathgoodies.com/lessons/vol5/challenge_vol5.html).

1. Mt. Everest, the highest elevation in Asia, is 29,028 feet above sea level. The Dead Sea, the lowest elevation, is 1,312 feet below sea level. What is the **difference** between these two elevations? (*subtract*)

2. In Buffalo, New York, the temperature was -14°F in the morning. If the temperature **dropped** 7°F, what is the temperature now? (*subtract*)

3. A submarine was situated 800 feet below sea level. If it **ascends** 250 feet, what is its new position? (*add*)

4. A submarine was situated 450 feet below sea level. If it **descends** 300 feet, what is its new position? (*subtract*)

5. In the Sahara Desert one day it was 136 F. In the Gobi Desert a temperature of -50 F was recorded. What is the **difference** between these two temperatures? (*subtract*)

www.ingramcontent.com/pod-product-compliance
Lightning Source LLC
Chambersburg PA
CBHW071214210526
45170CB00024BB/1402